Textiles
through the ages

Ruth Barnes
Emma Dick
Jon Thompson

Ashmolean Museum · Oxford · 2002

Introduction

This booklet is intended to illustrate the history of textiles and weaving through the centuries by bringing together some key objects from the Ashmolean Museum's many collections. The oldest objects tell us about the early history of the fibres used by weavers and how different textile techniques developed and spread in the ancient world. The later objects—not arranged in any strict sequence—show how textiles were used in society to indicate rank and status, and how they were transported from one culture to another, sometimes over great distances, to become both functional and luxury items, or occasionally exotic articles of high fashion.

The objects chosen include representations of textiles in a variety of media. The idea here is to invite people to walk through the Museum's galleries and take a fresh look at objects that may already be familiar, but this time from a different point of view. In addition to the textile representations, we have selected a number of actual textiles. These have been chosen in part to illustrate some point of special interest, but also to draw attention to the Museum's rich study collections. Of particular importance are the Newberry collection of Indian and Islamic textiles, the Shaw collection of Central Asian costume (both in the Department of Eastern Art) and the Mallett collection of English embroidery (in Western Art).

Embroideries in the Mallett collection (apart from the casket) are housed in locked drawers in the Founder's Room, while textiles in the Shaw and Newberry collections are kept in the Eastern Art Print Room. Access to all three collections is by prior appointment. Most of the other objects illustrated are on display; the more delicate items, however, are only shown for limited periods, so it will be a matter of chance as to how many objects in the booklet will be visible at any one time. Not adequately represented here, for reasons of space, are the Museum's collection of European tapestries. These are on permanent display in the Mallett Gallery, also known as the Tapestry Gallery.

We thank the following for their assistance: Mary Brooks, Catherine Casley, Julie Clements, Oliver Impey, Janice Katz, Arthur MacGregor, Roger Moorey, Canon Brian Mountford, Alex Newson, Phyllis Nye, the Photography Department of the Ashmolean Museum, Andrew Topsfield, Shelagh Vainker, Michael Vickers, Catherine Whistler, Helen Whitehouse, Jon Whiteley.

Note: Museum accession numbers are given in brackets following each caption.

Fragmentary painted sandstone statue of Akhenaten, 14th century BC, Egypt. (AN 1924.162)

The flowing garments draped around Akhenaten's some-what effeminate form would have been made of linen. The cultivation, processing and weaving of flax was highly developed in ancient Egypt. Sheer linens of exquisite delicacy have been recovered from wealthy burials as far back as the third millennium BC. The very fine threads required to weave such garments were obtained by splicing together individual fibres as they were spun. Linen was used for weaving long before wool and cotton and in Europe even antedates the appearance of pottery.

Limestone male figure, mid-3rd millennium BC, Iraq. (AN 1919.65)

There is scholarly debate as to whether the garment depicted on this figure represents a fleece or a piled fabric woven in imitation of one. Sumerian is rich in textile terms and there is some evidence that contemporary piled fabrics did exist; Leonard Woolley, excavating at Ur in the 1920s, reported having found a textile which had on one side of it what he described as a deep pile or tassels. Unfortunately it rapidly disintegrated on exposure to air and we do not know what materials the weaver used.

Moulded terracotta plaque of a *Yakshi* or mother goddess, c. 200 BC, India. (EA X.201)

The closely draped garment of this figure suggests a fine muslin for which India has long been famous, though it could have been an imported silk or possibly even wild silk. The epicentre of cotton in the Old World appears to be India, but precisely how old its use is remains a mystery. Fragments of a cotton textile have been found at Mohenjo Daro dating to 2700 BC, and other archaeological finds could be even earlier. Why there appears to be no mention of cotton in the Vedas remains unexplained. The incredibly fine cotton yarns used in muslins were spun using a tiny wooden spindle resting in a small bowl.

Boeotian black-figure pottery drinking vessel (*skyphos*) with a depiction of Odysseus and Circe on one side, 4th century BC, Greece. (AN G.249)

Clay loom-weights, 8th century AD, Middle Aston, England. (AN 1923.261)

The beautiful sorceress Circe was sitting in her palace singing and weaving a robe so fine as only a celestial woman could produce, when Odysseus' companions arrived. She offered them food and drugged wine; she then transformed them into pigs, and it was up to Odysseus to rescue them.

Circe is shown here with her upright, warp-weighted loom. Small ring-shaped weights of clay or stone were tied to the ends of the warp threads to keep them evenly taut. Clay or stone weights survive from early Neolithic sites, from the sixth or possibly even late seventh millennium BC. When these are laid out in a regular row they are likely to have been part of a warp-weighted loom.

This type of vertical loom appears in representations from the late Bronze Age onwards, from c. 1500 BC. The most detailed early images come from the black- and red-figure vases of Classical Greece. The warp-weighted loom was used north of the Mediterranean from Greece to Central Europe, and it spread eventually to Scandinavia and Britain, where it remained in use until the Norman conquest, as the loom-weights from 8th century England (right) illustrate. It still survives in Lapland to this day. Weaving on a warp-weighted loom starts at the top and progresses downwards, with the weft beaten in with an upwards motion.

Two fragment of tapestry-woven cloth, wool and linen, 4th century AD (above) and 6th–8th century AD (below), provenance unknown, found in Egypt. (AN 1933.506, AN 1939.445)

The technique of tapestry weaving—the use of discontinuous wefts to make the pattern—is both ancient and widespread and the dry climate of Egypt provides us with some of the oldest examples. They date from around 1400 BC and are worked in linen. These examples, woven with wool, are much later. Delicate examples of woollen tapestry are known from Achaemenid Persia (4th century BC) and coarser multi-coloured woollen tapestries, both abstract (above) and figural (below), are abundant in Late Roman burials in Egypt. They were used for clothing and in the home. The well-known kilims of Turkey and Iran are woven using the same technique.

Corner fragment of decorative linen cloth with looped pile formed by supplementary wool wefts, 4th–5th century AD, Egypt. (AN 1889.66a)

Archaeology has revealed that pile-woven textiles were widespread in former times and that many different techniques were used to make them. Looped pile was used in Pharaonic Egypt as early as 1900 BC and knotted pile around 1400 BC. Piled weavings with elaborate designs worked in wool of many colours became popular in late Roman Egypt. The pile of these 'Coptic' weavings was usually formed by inserting extra wefts as a series of loops, which is the technique used for this fragment.

The ground weave of this textile is linen. Contrary to popular belief cotton cultivation in Lower Egypt only began in the middle of the eleventh century and became widespread in the thirteenth and fourteenth centuries. Cotton was, however, extensively traded in the Mediterranean region and appears in Egyptian textiles well before it was first cultivated there.

Three fragments of weft-faced compound weave, wool and linen, 4th – 5th century AD, Egypt. (AN 1891.288, AN 1968.561, AN 1888.743)

Though small and relatively unimpressive, these fragments mark the beginnings of what was to become a major development in weaving technology. The great flowering of silk-weaving in the Near East—in Sassanian Iran (AD c.224–651) and the Byzantine world (AD c.364–1453)—is thought to have begun when silk imported from China was substituted for wool in weft-faced compound weaves such as these. Byzantine silks are as much as 2.7 metres wide and would have been woven on large, probably vertical, looms. Such looms still survive in a small town in Iran where blue and white cotton floor coverings (*zilus*) are woven for use in mosques. The semi-mechanical patterning system used in these looms was the forerunner of the true draw-loom which was developed around the beginning of the second millennium AD.

(right) Pall of Cloth of Gold, 15th century, Italy, on loan from the University Church of St Mary the Virgin, Oxford.

The term 'cloth of gold' is often thought to be a romantic fiction, but it does actually describe a type of textile woven with gold thread made from either thin strips of gilded parchment (or other animal membrane), or silk thread spiral-wrapped with gold or gilded silver foil. From the medieval period onwards richly patterned silks shimmering with gold were produced at great cost for the highest levels of society where their use signified a person's rank and status. This large pall consisting of four loom-widths of uncut, voided velvet is unusual in having both the pile and the ground worked in gold thread. It is probably of Italian workmanship and was given by King Henry VII to the University of Oxford in 1505, plus £10 per annum in perpetuity, providing that a service be held for him annually at St Mary's Church. The cloth was used as a pall to cover his coffin and was brought out each year at the anniversary of his burial.

Bodhisattva Guanyin **(merciful higher being), gilt polychrome wood, 10th century AD, China. (EA 1999.96)**

This rare wooden sculpture has remains of polychrome paint and fine details of patterning on its flowing robe. On the bodhisattva's shoulder a cloud design and a lotus can be identified, and several large lotus blossoms appear on the skirt. They have a raised surface to make them visible, but in the actual silk textile they probably would have been woven in a weft-faced compound twill (samite). The large lotus design appeared on Tang period (AD 608–918) textiles from the eighth century onwards, although the scrolling surround seen here is a characteristic of the 10th and 11th century Song style.

8

(left) Michele Giovanni Boni, called Giambono (1400–1462), *A Bishop Saint*, **Venice, egg tempera with gilding on wood. (WA 1984.149) [A1133]**

In the mid-fifteenth century silk weavers in Italy began to produce a new type of velvet with large-scale, bold patterns on a brocaded ground of gilded silver foil wrapped around a core of yellow silk. The effect was one of incomparable richness and splendour. The style was soon imitated in Spain and had a deep influence on the velvets of Ottoman Turkey. Many examples of such velvets have been preserved as ecclesiastical garments similar to the one being worn in this fifteenth century Italian painting of a 'Bishop Saint'.

(right) Leandro Bassano (1557–1622), *Portrait of a Procurator of Saint Mark*, **Venetian school, oil on canvas. (WA 1935.97) [A445]**

The textile covering the table in this painting is difficult to identify. At first sight it could be taken for a conventional rendering of a knotted-pile carpet. It is more likely, however, to be one of the much rarer silk tapestries which came from Iran to Europe by way of royal gifts and special commissions. This identification is made possible by some specific details of the border which can be matched with surviving examples. The identity of the man in the painting is unknown, however the crimson *stola* worn over his left shoulder, made of costly Italian voided velvet, and his gown, made of the same material, indicate his high rank. Other details of Venetian fashion, such as the height of his collar, the cut of his beard and his hair style suggest, according to expert opinion, that the painting dates from the 1590s. If that is the case, the type of silk tapestry on the table would have been produced earlier than is currently believed.

Jacopo da Ponte, called Bassano (Bassano del Grappa) (1510–1592), *Christ Among the Doctors,* **Italian School, c.1539, oil on canvas. (WA 1949.5) [A771]**

The almost photographic rendering of the fringe, the plain-woven end and the knotted details of the border make it possible to identify the carpet in this painting as Turkish. Carpets of this type were brought from Turkey by Venetian traders and became popular in Europe during the second half of the fifteenth century, remaining in fashion well into the sixteenth. The evidence we have for this is their appearance in Italian paintings in the 1450s and their disappearance in the 1550s. Although knotted-pile carpets with closely related patterns are known, no example with this precise border design survives.

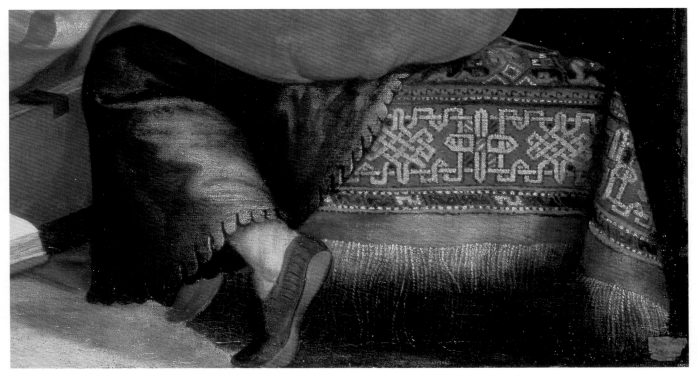

Simon Kick (1603–1652), *A Lady Seated at a Table***, Dutch School, early 17th century, oil on panel. (WA 1851.21) [A124]**

Paintings of Dutch interiors frequently depict carpets covering tables. The accuracy with which this design has been rendered makes it possible to identify the carpet precisely as a type known as 'chequerboard', so called because of the division of the field design visually into squares. Such carpets must have been exported in quite large numbers, though their country of origin is not definitely known; current scholarship favours Syria. The same carpet appears in another painting by Simon Kick in the National Gallery of Ireland, Dublin.

Two block-printed cotton fragments, early 14th century, India, made for export to Egypt. (Newberry Collection, EA 1990.807, EA 1990.973)

We know from historical sources that in medieval times Indian cotton fabrics were exported to countries on every shore of the Indian Ocean. Many of them have survived in Egypt, and the Ashmolean has the most important study collection of these early, mainly block-printed textiles in any public museum worldwide. More than 1200 pieces were donated in 1941 by the Egyptologist P. E. Newberry, together with an equally impressive collection of medieval Islamic embroideries (see below). The Indian textiles were used for garments and furnishings. As radiocarbon analysis of key pieces has shown, most of the collection dates from the tenth to the fifteenth century. At that time Indian craftsmen were unrivalled in their mastery of dyeing cotton, a fibre which does not absorb colour easily. They excelled particularly in the use of indigo and red dyes. The outlines of the small geese in one of the fragments were first printed in a resist. Then different mordant substances were applied prior to the red dye bath, thereby producing distinct shades of red, brown, and pink. This fragment, along with several others in the collection, provides primary evidence for the global nature of the Indian textile trade, as identical cloths were also traded to Indonesia in the fourteenth and fifteenth century, where they have survived on the island of Sulawesi.

Linen garment or scarf fragment embroidered with silk, Mamluk period, radiocarbon dated 1395 AD +/− 40, Egypt. (Newberry Collection, EA 1984.445a)

Although now fragile and fragmentary, this is an outstanding example of Islamic medieval embroidery. Like the Indian textiles previously described, it is part of an impressive collection of textiles donated to the Ashmolean by the Egyptologist P. E. Newberry. There are more than 1000 medieval Islamic embroideries, most of them probably made in Egypt. The collection is unique in size and variety, with many pieces of exceptional quality, as this fragment shows. It was patterned in very fine pulled and drawn work, embroidered in combination with double running stitches for outlines and counted satin stitches to fill in. The different embroidery techniques used side by side provide a particularly successful contrast of texture. Good quality embroideries survive from the Mamluk period in Egypt (1250–1516), and their geometric designs occasionally appear in Italian Renaissance painting. Eventually they were copied in sampler books and by the seventeenth century were adopted in European embroideries. Highlights from the collection are published in Marianne Ellis' book *Embroideries and samplers from Islamic Egypt* (2001), published by the Ashmolean Museum.

**Silk lady's robe
kosode, 18th
century, Japan.
(EA 1989.203)**

In Japan garments are
often classified by
their types of sleeve
(*sode*). Robes from
the Edo period (1615–
1868) generally had a
small wrist opening
and are therefore
called *kosode* (literally
'small sleeves').
Decoratively, there
was a shift away from
the overall patterned
designs of earlier
periods to a more
restrained use of
ornament, as appears
here. The fine satin
stitch embroidery is
combined with
stamped gold and
silver foil (*surihaku*) on
a yellow satin-weave
ground. The motifs of
the phoenix and the
branch of the
Paulownia tree are
both considered
auspicious; according
to legend the
migrating phoenix only
rested in Paulownia
trees.

Embroidered casket, c.1665, London. (Mallett Bequest, WA 1947.315)

Embroidery was an important part of a girl's education in the seventeenth century, as the exceptionally good documentation for this casket shows. An eighteenth century letter in one of its drawers says, 'The casket was made by my Mother's Grandmother who was educated at Hackney School after the plague in London all the young ladies works were (burnt) [crossed out] destroyed that they were about at that time, she left school soon after therefore this was made viz. before 1665.'

The Ashmolean Museum has a fine collection of seventeenth century three-dimensional raised embroidery, so-called stump work, which was fashionable in England at the time. It was often used to depict biblical narrative scenes. The sources for the narratives were engravings that circulated widely. Thus certain settings and iconographic characteristics are repeated in different embroidered pictures. A favourite was the story of Abraham and Isaac. On this casket Abraham's wife Sarah and their son Isaac appear on the lid, while the two doors on the front show Abraham's dismissal of his concubine Hagar (left and detail) and Hagar and her son Ismael in the wilderness (right).

The right side panel has Abraham sending his steward Eliezer to find a wife for Isaac; the story continues on the back with Rebecca at the well offering water to Eliezer, and concludes on the left side panel, where Isaac and Rebecca are shown as a betrothed couple. The narrative is intended to celebrate marital fidelity and a wife's loyalty to her husband.

The sides and back of the box are embroidered in simple tent stitch and were obviously not intended as the primary focus of the piece, while the front and lid are exquisitely decorated with complex layers of raised work. Abraham and Hagar's costumes are worked in needlepoint lace using different fibres and sizes of loops. The resulting effect is very successful, with the different textures of their dress clearly distinguished. The effect of coiffure is achieved with coiled metal thread, possibly wrapped with human hair. Peacock feathers were originally used to give an iridescent shimmer to a caterpunate's body (unfortunately only left in minute traces), and the windows of the house are made of mica.

Needlework hanging, *A Musical Party***, mid-17th century, Spain or Portugal. (WA 1991.180)**

Although at first glance this large (approximately 237 x 407 cm) wall-hanging may look like a tapestry, it is in fact an embroidery on linen. The ground fabric is in wide horizontal bands joined together with invisible flat seams and embroidered in silk and metal threads worked mainly in Gobelin stitch. It would have been stretched over a frame and rolled up as it was embroidered, which explains the somewhat distorted perspective and proportions of the scene. This piece came from a country house in Staffordshire where it fitted exactly the space in which it was displayed, so may have been cut down. Similar pieces sometimes have borders bearing Spanish coats of arms. Large needlework "carpets", worked on linen ground are known from the town of Arraiolos in Portugal. Whether the piece was made in imitation of a tapestry, or conceived initially as an embroidery may provide clues as to its provenance. Stylistically, it seems to follow northern European sources, though technically it fits rather better with a southern European manufacture. The most common stitch in largescale northern European needlework is the tent stitch, while the Gobelin stitch used in this piece is usually found in large hangings from Italy, Spain or Portugal, though the Italian pieces tend to be more painterly. For the present the hanging's origin remains somewhat enigmatic.

(left) Attributed to Marcus Gheeraerts the Younger (1561–1636), *Portrait of a Lady, possibly Lady Elizabeth Poulett*, **Anglo-Dutch school, oil on canvas. (WA 1845.2) [A2]**

This striking portrait is remarkable for the detailed depiction of what a wealthy and fashionable woman would be wearing in the early seventeenth century. Her identity is uncertain, though she is believed to be Lady Elizabeth Powlett or Poulett, the wife of John, first Baron Powlett of Hinton St George, Somerset. The swirling pattern of her embroidered jacket was a popular design at the time and excellent examples of actual garments in this style are preserved in the Burrell collection, Glasgow and the Victoria & Albert Museum, London. Also of interest are her lace collar, cuffs and apron, the result of hundreds of hours of work, and the gold-trimmed bands, which were probably tablet-woven. Unfortunately the picture is in such fragile condition that until the specialised conservation work it needs can be completed it must not be moved from its storage bay, let alone exhibited.

Embroidered stomacher, 18th century, England. (AN 1921.854)

In the eighteenth century the pursuit of fashionable elegance involved wearing this item of clothing known as a stomacher. It was tied around the body as a separate panel with tapes passing through the looped tabs on both sides. It was worn under a gown with an opening at the front designed to reveal the decorative stomacher and petticoat beneath. On formal occasions it could be stiffened with an insert—the busk—made of wood, whale bone, metal or other material. The medieval collection has a large, decorated, shield-shaped busk made of tortoiseshell. The overall effect was to emphasise a narrow waist, but in doing so it prevented the unfortunate wearer from bending forward. The silken laces look functional but are purely decorative.

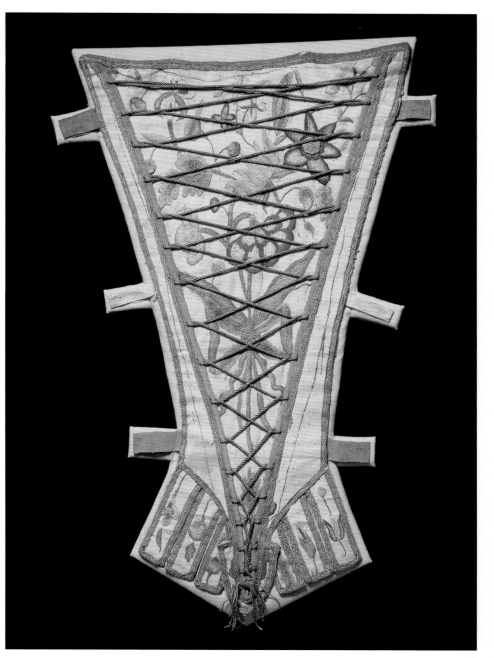

Silk *oshi-e* screen (detail, third panel), 1893, Japan. (EA 1995.87)

The four-panelled screen is covered with fabric and worked in a Japanese version of raised technique called *oshi-e*, where silk or cotton padding is wrapped with cloth and pasted onto the surface. Here the figural scenes are further enhanced with embroidery. Making *oshi-e* became a favourite pastime for women of aristocratic and warrior families in the seventeenth century. This screen, though, is much later and was made professionally. The topic represented – the four officially recognised classes of samurai, farmers, craftsmen, and merchants – gives a panoramic view of Japanese society. The subject matter, combined with the quality of workmanship and the unusually heavy hardwood frame suggest that this screen was made for one of the great international exhibitions of the second half of the nineteenth century. As one of the figures carries a bundle with the date 1893, it may have been intended for the World Exhibition held in that year in Chicago.

   The third panel from right, representing Japanese crafts, has three scenes that relate specifically to textile production. Placed prominently in the centre of the scene is a large draw-loom, showing the weaver seated at the cloth beam, pulling the reed towards himself, while the draw-boy crouches above the far end of the loom to manipulate the so-called lashes that control the warp ends and make it possible to weave figured textiles. Just below the loom is a man working on an embroidered panel, while above the weaver is a seated woman making braids.

(right) Man's silk *ikat* robe, pre-1869, Central Asia, Chinese Turkestan. (Robert Shaw Collection, EA X.3977)

The coat is one of several garments given to the English explorer Robert Shaw when he travelled to Chinese Turkestan in the autumn and winter of 1868/69. He was the first Englishman to reach Kashgar and Yarkand, and his diary (published in 1871 as *Visits to High Tartary, Yarkand and Kashgar*) gives a vivid account of people's lives, including how they dressed. Shaw was quick to note that the robes he received were indicators of prestige. This coat with its many colours was clearly in the top rank. On several occasions Shaw records the date and location of when he was given garments, some of which can still be identified. This makes his collection not only one of the earliest recorded in the nineteenth century, but also one of the best documented.

   The robe is decorated with *ikat*, a resist-dye technique that is particularly well developed in several parts of Asia. *Ikat* comes from the Malay/Indonesian word for 'to tie'. The warp or weft threads are tied with a design and are dyed prior to weaving the cloth. The technique depends entirely on the coordination of hand-tied knots and cannot be produced with the help of mechanical devices. At present it is not known when and where the technique first developed. It may be detected in dress shown in the Ajanta cave paintings of India (fifth to seventh century), and a seventh century ikat fragment in the Shosoin shrine at Nara, Japan was probably produced in Central Asia. Maritime South-east Asia is another contender for the origin of the technique. Ikat textiles also were made in the Yemen by the tenth century and were traded to Egypt, from where fragments have survived.

Detail of double *ikat* silk *patolu*, 19th century, Gujarat, India. (EA 1962.13)

A particularly taxing version is double-ikat, where both warp and weft are tied and dyed, and the design is then matched up during weaving. This detail shows how a *patolu* (pl. *patola*), one of the most spectacular Indian silk ikat cloths, was patterned. Prior to weaving, the design was tied as a resist into both warp and weft, and the threads were then dyed, a process that had to be repeated for different colours. The warp was set up on the loom, and the weft inserted to exactly match the pattern. As warp and weft combine to show the design, they have to be evenly spaced, and therefore a balanced plain weave is used. *Patola* have a long and distinguished history, both in India and abroad. They were much in demand in South-east Asia even before Europeans arrived in the region in the early sixteenth century, and their designs became an inspiration to local weavers. Although now only one family workshop in Patan (Gujarat) produces genuine silk *patola*, they are still made to order and are used by wealthy Indian families as wedding saris.

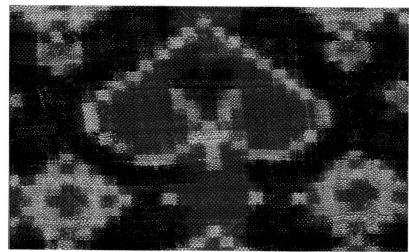

Maharajah Ajit Singh of Jodhpur, c. 1720, Jodhpur, Rajasthan, gouache on paper. (EA 1990.1283)

(right) Court girdle *patka* in silk with silver-gilt thread, 18th century Mughal India. (EA 1962.13)

The Indian painting shows the ruler of Jodhpur wearing an outfit that is both sumptuous and restrained. His red coat has an overall golden pattern of small floral clusters, possibly representing gold thread embroidery. His white turban cloth is embellished with a pale green pattern that may be tapestry-woven or embroidered. Around his waist he wears a girdle called a *patka* which has mauve-coloured ends and a wide border with a row of Mughal-style gold and green flowers. It would have been woven in silk brocading against a gold metal thread background, as can be seen in the detail of the actual *patka* border (right). This has a main body of apricot-coloured silk woven in plain weave, but at both ends the colour of the weft changes to dark blue and a gold metal thread supplementary weft is added. The red, green, and white parts of the flowers are brocaded, i.e. inserted as discontinuous supplementary weft.

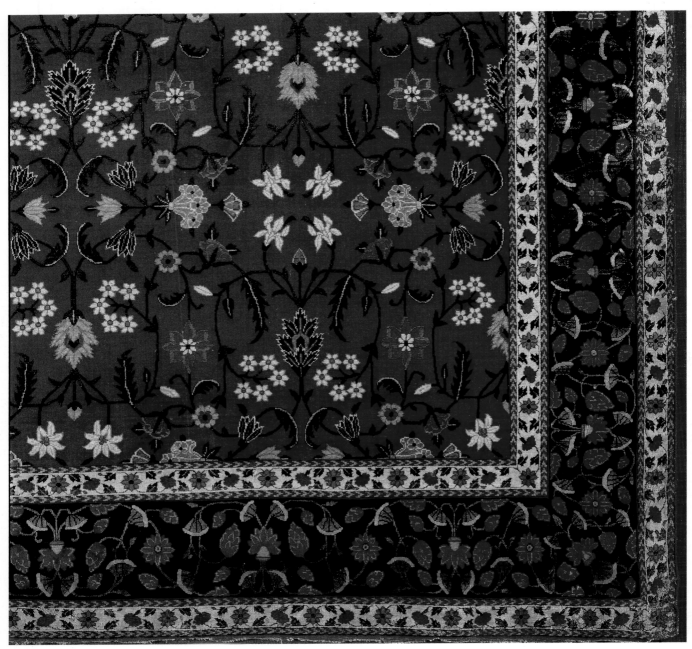